YOUR KNOWLEDGE HAS VALUE

- We will publish your bachelor's and
  master's thesis, essays and papers

- Your own eBook and book -
  sold worldwide in all relevant shops

- Earn money with each sale

Upload your text at www.GRIN.com
and publish for free

The Impact of Climate Variables on Food Western Rajasthan Security. A Case Study in Jodhpur District

Asif Iqbal

Bibliographic information published by the German National Library:

The German National Library lists this publication in the National Bibliography; detailed bibliographic data are available on the Internet at http://dnb.dnb.de.

ISBN: 9783346713896
This book is also available as an ebook.

© GRIN Publishing GmbH
Nymphenburger Straße 86
80636 München

Print and binding: Books on Demand GmbH, Norderstedt, Germany
Printed on acid-free paper from responsible sources.

The present work has been carefully prepared. Nevertheless, authors and publishers do not incur liability for the correctness of information, notes, links and advice as well as any printing errors.

GRIN web shop: https://www.grin.com/document/1268909

SUBMITTED FOR THE AWARD OF THE DEGREE OF

MASTERS IN GEOGRAPHY

By

NAME – ASIF IQBAL

DATED – 25TH JULY 2022

ASSOCIATE PROFESSOR

CENTRE FOR THE STUDY OF REGIONAL DEVELOPMENT

SCHOOL OF SOCIAL SCIENCES

JAWAHARLAL NEHRU UNIVERSITY

CONTENT

INTRODUCTION

Agriculture usually plays a larger role in developing countries than in developed countries. For example, Indian agriculture accounts for about 20% of GDP and provides nearly 52% of employment (compared to 1% of US GDP and 2% of employment). Most agricultural laborers come from the social class. (FAO, 2006). In addition, farmers in developing countries are expected to find it harder to adapt to climate change due to credit constraints and limited access to adaptive technologies. However, most of the literature on climate change impacts focuses on developed countries, particularly the United States, probably due to data availability. Most studies in developing countries found surprisingly large potential impacts following a production function approach (Cruz et al., 2007). It is difficult to conduct a true Ricardo study in a developing country context because land markets are unlikely to function well and data on land prices are not widely available. Instead, the Semi-Record approach used data with average incomes, rather than the idea that if land prices were available, they would be the discounted value of current incomes. Large Ricardo studies in developing countries in India and Brazil were outranked by mild climate change scenarios (up to 2.00°C in mean temperature and 7% in precipitation). We found a severe adverse effect leading to 10% of agricultural profit (Sanghi et al., 1998b, 1997).

This study uses a panel approach to food security in Rajasthan using a panel in Jodhpur district covering the period 1960-1991. Annual climate data (temperature and precipitation) and yields for close canton effects. The resulting estimate of the meteorological parameter is then identified from district-specific deviations of the annual meteorological from the average district climate. Because annual climate change is inherently random and therefore unrelated to other unobserved determinants of agricultural outcomes, these panel estimates suffer from deductible problems associated with the hedonic approach. It should not be. The use of district-level data is important to obtain a reasonable change in climate over the year, thereby differentiating climate impacts from other annual shocks at the national level. We also include time trends in smooth regions to avoid confusing the effects of the slowly warming climate of the second half of the 20th century with the increase in agricultural productivity over the same period. The predicted average impact of climate change is then calculated as a linear combination of estimated meteorological parameters and predicted climate change.

The purpose of this review is to provide a critical review of the current extensive literature on the closely related relationship between climate change and the food system. In particular, it focuses on the broader issues of the food system beyond food production, highlights the distribution of climate-related food security impacts across different sectors of global society, and offers opportunities for food system integration. We aim to emphasize the challenges. And options for mitigation, adaptation, and food security.

Observed and predicted climate change drivers and patterns are well documented. The same radiative forcing drives both, but it is possible to distinguish between the long-term (10-year) trend and the short-term increase in climate change. In the long run, without complete mitigation measures, society will have to adapt to gradual changes in temperature and precipitation averages and distributions. Depending on the speed and direction of these trends, gradual or transformative adjustments are needed. Most immediately, climate change is perceived as an increase in temporal and spatial variability in temperature, precipitation, wind, and in particular the frequency and magnitude of extreme events. Types of extreme events that are expected to increase include the frequency and intensity of heat waves, the frequency of heavy rains and associated floods, the intensity of tropical cyclones, and the occurrence of very high sea levels due to storm surges.

AREA OF THE STUDY

Rajasthan, located in the western part of India, faces severe water shortages, has low rainfall, and is called a dry/semi-arid region. Officially, the state includes 33 locales, 39753 occupied cities 249 Panchayat Samities, and 9168 Gram Panchayats. Geographically, the state desert covers a whopping segment of land. The total mass of the state is about 56.5 million. The state's population of is 165 per square kilometer. km. In addition, it varies from 13 per square kilometer. In the Jaisalmer area, up to 471 per square kilometer.

The Jodhpur district, one of the largest locations in Rajasthan, is organized primarily in the western part of the State. This area ranges from latitude 26 ° 37'north latitude to 72 ° 55'east longitude 73 ° 52'east longitude. The surface is in the shape of an unpredictable square, like a sneaker, with its large sides slightly dented. The most exceptional length of the region from north to south is, which is 197 km, and its most striking width from east to west is 208 km. The area is oriented at an altitude of between 250 and 300 meters above sea level. It is adjacent to Rajasthan's five unique territories. It is surrounded by the Jaisalmer and Bikaner regions to the north, the Barmer and Pali regions to the south, the Nagaur and Pali districts to the east, and the Jaisalmer region to the west. The land area of this area is 22,850 km², which is the sixth. 68% of the total geological areas of the state. (Fourth location of the state) 22641.440 km² states (99.09%) and 208.60 km² urban zone (0.91%). This area belongs to the arid region of Rajasthan. Covers 11.60% of all areas of the state's bone arid zone.

OBJECTIVES

(1) To recognize the effect of environmental change on characteristic asset accessibility in the territory.

(2) To survey the effect of environmental change on nourishment security, work and destitution.

(3) To propose adjustment and eco framework versatility techniques including characteristic assets the board and harvest improvement.

(4) To inspect approach choices and institutional setup to guarantee to empower situations to adapt to environmental change impacts.

(5) To identify the impact of climate change on natural resources available in area.

(6) To assess the impact of climate change on food security, livelihood and poverty.

(7) To propose adaptation and ecosystem resilience strategies including natural resources management and crop improvement.

(8) To examine policy options and institutional setup to ensure enabling environments to cope with climate change impacts.

METHODOLOGY

The purpose of this study was to take into account the size of the current problem, encounters in other helpless areas, the standard of activity, and the early effects of environmental changes on food security in the western region. Is to provide a rating. The data collected at Inquiry laid the foundation for the proposed 2010-2016 Food Security Action Plan. Of the networks and governments, respond to the impact of environmental change on food security. The basic purpose of this audit study is to:

(I) PARTICIPATORY RESEARCH APPROACH-

The participatory research approach entered the field of improvement. These methodologies are also called participatory rural assessments. This methodology involves locals who are interested in organizing, reviewing, and rating promotional programs.

(II) <u>HOUSEHOLD REVIEW -</u>

The participatory research approach entered the field of improvement. These methodologies are also called participatory rural assessments. This methodology involves neighbors who are truly interested in organizing, reviewing, and rating promotional programs.

(III) <u>EXPERT INTERVIEWS -</u>

This kind of discussion takes place with various experts. National, state, and local level people.

(IV) <u>STATISTICAL METHOD -</u>

Mean, mode, median, Co-relation, SD (Standard Deviation), regression, t-test, f-test, climate variability, rainfall trend analysis, uncertainties, etc.

(V) <u>CARTOGRAPHY-</u>

Map, Diagrams, Photographs & other analytical methods.

(VI) <u>FIELD VISIT –</u>

To understand the local community's perceptions and their cultural practices for proposing programmatically viable adaptation strategies, regret-free adaptation strategies were implemented in selected districts, concerning climate data/food production, etc. The number of stakeholder workshops for which secondary data is collected and analyzed to discuss and finalize the implementation results. Study scoping through village visits and stakeholder discussions. Field Test development and completion of research equipment. Conducting recent surveys and tracking participation spending.

<u>HYPOTHESIS</u>

This study aims to find the following two questions about the current state of food security in western Rajasthan. This hypothesis may also form the basis of research on food security in western Rajasthan.

(1) The impact of variables of climate change has affected already stressed environments imparted social, geographical, demographic, and migration patterns and its concomitant response planning and has affected the people themselves, their cultivated land, livestock, social groups, and food security.

(2) There is a gap between the demands from social groups, stakeholders, and government intervention to identify the basic needs of food security.

<u>IMPORTANCE OF THE STUDY</u>

The importance of geographic research is limited to the identification and profiling of research institutes working on climate change and food security in Rajasthan. This topic does not cover the science of climate change and food security, as this topic is better covered elsewhere by geographic studies (social groups) in Area. Each of the four elements of food security, specifically accessibility, availability, use, and frame strength, is affected by changes in the environment. The assessment of the impact of environmental changes on the food and agricultural portion focused on the angle of access, especially the disappointment of crops and the reduction in crop yields. Aspects of food security accessibility and use depend on sociocultural conditions such as affordability, food preferences, and nutritional value. For a populous country like India, which is affected by climate change, the threat to food security is changing in many ways. An overall analysis of the food system will be carried out, along with a review of government policies to address this issue.

<u>**REVIEW OF LITERATURE**</u>

The audit covers books and looks into articles, reports, and mainstream articles on the exploration point. The primary subjects canvassed in the survey are Climate Change and Food security.

Olson CM (1996) reviewed advances in First World food security research over the last two decades, "when vibrant and delicious foods and safe foods became available. The state of that is in vogue. " The source or ability to obtain sufficient food in the socially commendable habit is obligatory or wrong. Food insecurity is distinguished at the level on an individual and family basis, and evidence is based on confirmation of the rating by use in the study. Based on methods such as Community Childhood, Hunger Identification Project, and Cornell / Radimer Scale Finally, food insecurity is considered for human and monetary social spending, suggesting the need for government mediation in society. I am. The problem is presented as an open problem.

Vyas VS (2000) conducted a study focused on the nutritional assessment of food consumption. It argued that if food intake meant food security, it would mean. Vyas ensured food security, focusing on the fact that it served functions of the state, showcase, and society, and that they corresponded to each other.

Dilly and Boudreau (2001) characterized the defencelessness of the food security situation, responding to consequences such as appetite and hunger. The safety of the family unit is estimated by fully practicing the methods available and determining whether it will receive adequate nutrition in all seasons for an additional year. The relationship between these options and the different stock variables determines the weakness of the family unit.

DISCUSSION

<u>**CLIMATE CHANGE**</u>

Rathore et al. (2006) In their article, point out the recognition made from the beginning of the 21st century and points out changes such as precipitation of, surface temperature, the rise of average sea level, the disappearance of, and abnormal events. It shows that. These recognitions were also made in India. The authors push the hotter atmosphere into the water cycle, adversely affecting the water table in the long run. The paper seeks to examine the potential impact of environmental changes on India's groundwater and the surface of the earth. Research relationships between food production, precipitation, freshwater demand, and population. When discussing surface water resources, waterways, holidays, and pieces of ice were mentioned. When examining surface waters, the impact of floods and dry season events should be considered. It has been pointed out that groundwater is being abused to meet evolving needs in homes and modern parts and is rapidly declining as it has become a basic supplier.

The atmosphere in the Jodhpur area is usually hot and semi-dry, but it is during the stormy season (Köppen BShw) from June to September. Despite the fact of that normal precipitation is about 450 millimeters (18 inches), it is an astonishing factor. In the year of famine in 1899, Jodhpur was only 24 millimeters (0.94 inches), 4, but in the year of the flood in 1917, it was 1,178 millimeters (46.4 inches). Temperatures from March to October are exceptional, except when the monsoon rains bring thick fog and the drops slightly. For long periods from April to June, maximum temperatures usually exceed 40 ° C. During the stormy season, the room temperature drops slightly. Nevertheless, the general low maggie in the city is increasing, which contributes to the typical anxiety of the warmth. Near Jodhpur, Phalodi is the driest place in the region as a state.

<u>**CLIMATE AND AVERAGE MONTHLY WEATHER IN JODHPUR**</u>

Jodhpur has a semi-dry atmosphere, is still hot, and is best visited from October to January 1 with a little downpour. Storm / Storm season is from June to September. The summer months are hot and instruct you to keep a strategic distance from Jodhpur during these months.

What is the best time of year to go to Jodhpur in India? Here are some average weather facts we collected from our historical climate data:

• During the long stretches of January, February, November &December you are on your way to experiencing a great climate with amazing normal temperatures.

• Hot season/summer is in March, April, May, June, July, August, September, October, and November.

• March to November is hot weather/summer.

• Most of the rainfall (misty season) occurs in July and August.

• Jodhpur has a dry period from January to May and October to December.

• The hottest month is May on normal.

• The coolest month is January on normal.

• July is the hottest month. This month should be kept away from the event that you don't care about too much fall.

• The driest month of March.

TABLE 1: TEMPERATURE AND PRECIPITATION PER MONTH OF JODHPUR DISTRICT

Months	Temperature			Precipitation
	Normal	Warmest	Coldest	Normal
January	14.3°C	23.0°C	5.6°C	0
February	17.1°C	25.5°C	8.8°C	1
March	23.4°C	31.8°C	15.0°C	1
April	30.2°C	38.2°C	22.1°C	0
May	34.3°C	41.7°C	26.8°C	2
June	35.2°C	41.6°C	28.8°C	3
July	32.8°C	37.8°C	27.7°C	6
August	31.7°C	36.6°C	26.8°C	5
September	30.7°C	36.7°C	24.7°C	3
October	27.7°C	36.2°C	19.1°C	0
November	21.5°C	30.7°C	12.1°C	0
December	16.1°C	25.3°C	6.9°C	0

FIG. 2: TEMPERATURE AND PRECIPITATION PER MONTH OF JODHPUR DISTRICT

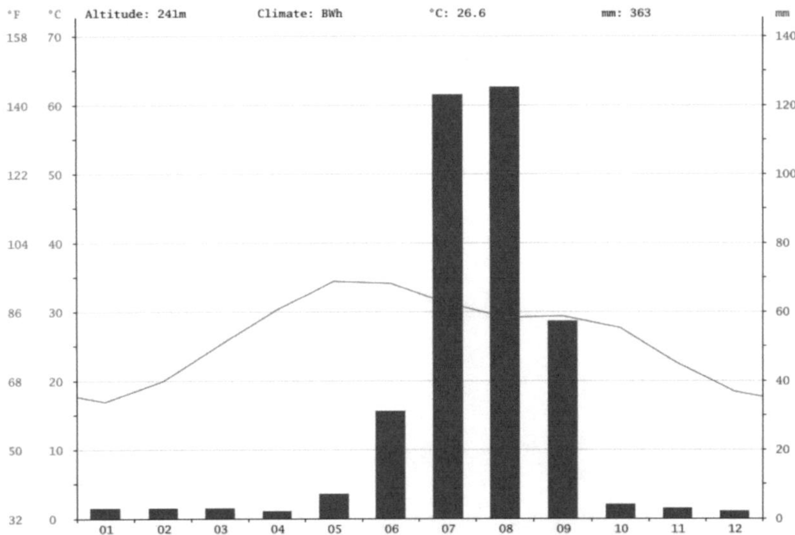

AGRO-CLIMATIC REGION

The climatic conditions of the district affect the horticultural design and different regions of yield different yields. Within a large group of climatic elements, precipitation, temperature, humidity, wind speed, sunlight length, etc. have a significant impact on the editorial design. Specific components that affect personal gardening and lifestyle. Factors such as temperature, rainfall, latitude, elevation, soil, yield, natural vegetation, and livestock are taken into account when choosing a horticultural district in Rajasthan. In 1980, the ICAR Research Review Board established rainfall, soil type, irrigation accessibility, water, existing management design, and authority, to distinguish between obvious problems in the area and their responses. Based on a unit, the state was characterized by nine agricultural climate locales. Later, another agricultural climate site was constructed during the reorganization of the Arid Western Zone in 1996. As of, there are agricultural climate areas in the state. A brief description of these locations is as follows:

ZONE-WISE DETAILS OF THE AGRO-CLIMATIC REGION OF JODHPUR

ZONE	RAINFALL	MAJOR CROPS	TYPES OF SOIL	DISTRICTS
IA & Western	200 - 370	In artificial rainfall, rain crops such as vajra, Kharif, legumes, and guar are mainly cultivated, and mustard is cultivated only in areas where irrigation water is available.	Floating soil in desert soil and dunes, coarse sand structure, calcareous in some places.	Jodhpur

RAJASTHAN AGRO-CLIMATIC ZONES

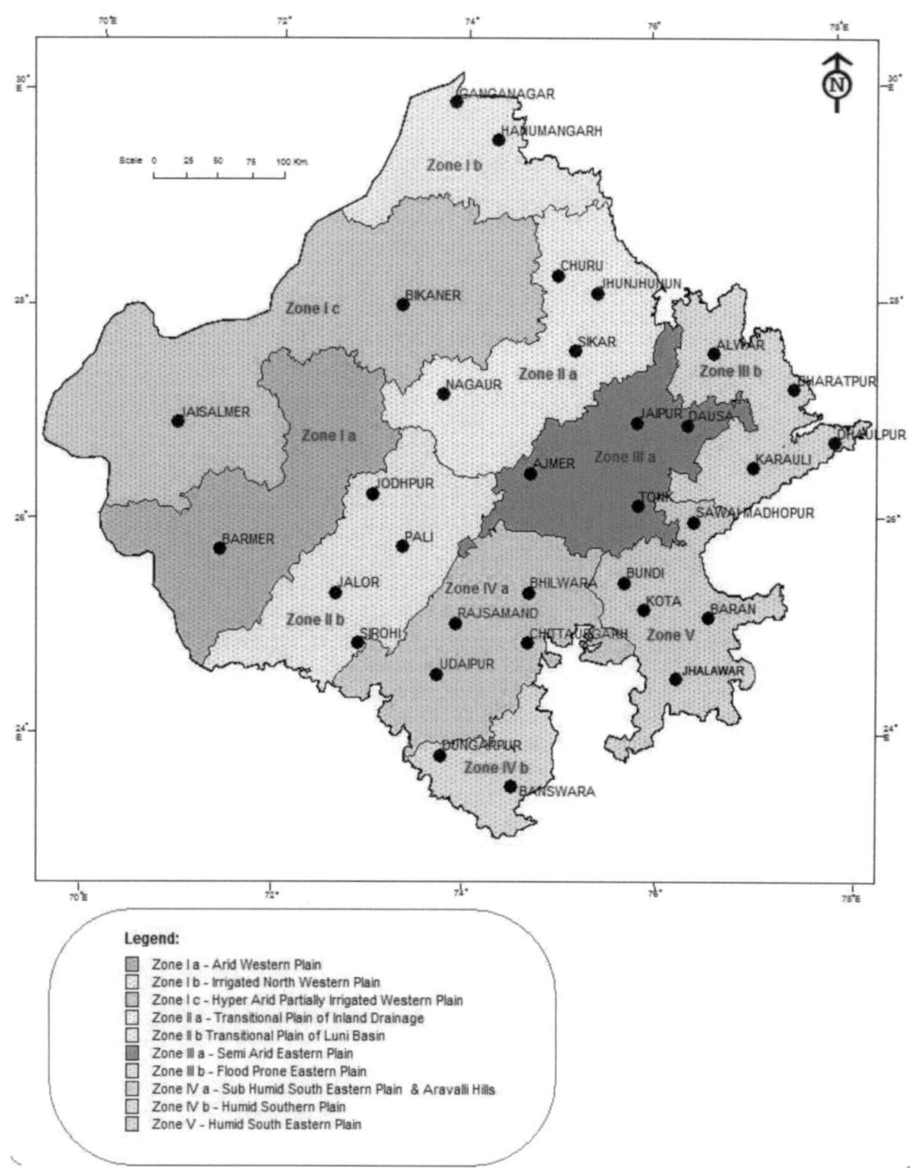

SOURCE: SOIL RESOURCE ATLAS OF RAJASTHAN, 2010

CLIMATE

Around Jodhpur, low rainfall is dry with whimsical vehicles, and daytime and annual temperatures are limited to, less haze and higher wind speeds. Rajasthan cannot see widespread atmospheric outbreaks from the modified geology. Very dry, wide fluctuations and temperature limits, and irregular and suspicious precipitation represent the environment in the region. Winter is from November to March, and late spring is from April to June. The period from July to September 9444 defines the southwestern storm season, and the post-storm season lasts from mid-September to October. The atmosphere in this area is dry, with an average annual rainfall of 366 mm, 425 mm to about 200 mm from southeast to northwest. The most intense is July, meaning Blustery days a year in the range of 14-21 during the June-September storm season. Precipitation variability is large, with a standard deviation of 123-226 mm and a coefficient of 49-55.

Looking at previous information, we can see that 81-150 percent of average annual rainfall, accounts for only 41-52 percent of the various stations in the region. Long-term under-rainfall (less than 80 percent of a typical) accounts for 31 to 22 percent. The most extreme temperatures in April, May, and June range from 38 ° C to 41.6 ° C, while the lowest temperatures in December and January range from 6 ° C to 10 ° C. There are some frozen waves and the probability of an ice event is once every three years. Normal wind speeds from March to June are 13.4 km / h for Jodhpur and 18.3 km / h for Phalodi. One day means that 20-30km / h is normal during this period. There are two meteorological observations in this area. One is Phalodi and Jodhpur, respectively, who talk about the climatic conditions of the regions in the northeast and southwest of the region.

RAINFALL

The normal annual rainfall in this area is 325.3 mm. with maximum incremental precipitation from northwest to southeast. Annual rainfall is almost 20 cm off. A little over 40cm at the north-western end. Located in the exceptional southeast of the district. About 80% of the rainfall falls during the rainy season in the southwest. The range of annual precipitation per year is enormous. During the 50 years from 1901 to 1950, the maximum annual rainfall in the region decreased from the usual in 1917 to 288%. The following year (1918) was associated with only 18% of the lowest rainfall. Typical. In the same 50 years, rainfall was below 80% of normal in 17 years. For the second consecutive year, there have been three events in the region, totaling, which is less than 80% of the normal rainfall.

TEMPERATURE

There are two meteorological stations in the area, one in Jodhpur and the other in Phalodi. In halo, temperatures are mostly cold in winter and warmer in the middle of the year than in Jodhpur. After March, the temperature will rise sharply. May is the toughest month in Jodhpur and Phalodi, with an average daily maximum temperature of 41.6 ° C. Daytime temperatures in June are slightly lower, but nighttime temperatures are higher than in May. The last few months of spring are, which is very hot with a scorching breeze. The most extreme temperatures can exceed 47 ° C. Since mid-July, temperatures appear to be declining as thunderstorms flow into the area. After the southwest storm has passed, the daily highs for options will be reached in October. From this point on, both daytime and nighttime temperatures of drop rapidly. January is the coldest month in Jodhpur, where the average daily temperature is 9.5 ° C, while in Phalodi it is 6.6 ° C.

Throughout the winter, cold waves affect the region after has passed the disturbing effects of the west, with soil temperatures sometimes dropping from 2 or 3 ° C below freezing water. The daily temperature difference is very large, and the temperature drops after dark very quickly during the winter.

The terrible cold wave of downtime and the warmth of summer are attributes of the desert. Temperatures in western Rajasthan begin to rise stoutly from March to June. In areas like Jodhpur, Jaisalmer, Barmer, and Bikaner, diurnal temperatures vary most extreme from 40 ° C and 45 ° C. It sporadically rises to 49 ° C between the extended months of May and June. In the summer, the diurnal temperature range becomes decreasingly important. The base temperature in the summer gradationally drops at night, staying at 20 ° C to 29 °C. Notable corridors of the state, including the Western thirsty and Midwest Semi-Arid, have normal veritably extravagant temperatures of 45 °C. December and January are the downtime months with the least temperature drops.

IMPACT OF CLIMATE VARIABILITY

Global climate change is affecting many parts of the world, leading to water scarcity and desertification (Houérou, 1996; Sellers et.al., 2008; Costa and Soares, 2012; Hillel. and Rosenzweig, 2002; Sivakumar, 2007; Odorico et Al., 2012) It will be necessary to study the climatic conditions of the already stressed arid regions of the world. A handful of available literature on Rajasthan's climate studies does not provide conclusions about the generally accepted temperature and precipitation conditions. An ancient study by Banerjee and Upadhya (1975), used climate data from 1869 to 1969 to examine the magnitude of drought in the region. Tendency to increase the severity of drought. Pant and Hingane (1987) were interested in the annual surface temperature and precipitation pattern in the region of north-western Rajasthan, India. During the study period of frames (1901-1982), they conducted a study conducted by Winstanley (1973), in which the annual surface temperature decreased and the monsoon precipitation conditions increased each year, as well. I concluded that I had defeated it. Rathore (2005) investigated the frequency of the dry season over the past few years and found that extreme and very severe dry seasons occur in states in the south and west, with a 47% chance of. .. (Ranade et.al., 2008) examined environmental changes throughout India. From there, there was a decrease in annual rainfall patterns between 1976 and 2006. RAPCC (2010) also used the IMD (India Meteorological Department) 1 ° x1 ° grid data to investigate the frequency of the previous century's dry season and place areas that are increasingly vulnerable to the continuous dry season. did. The SPACC study also looked at glimpses of drought rainfall and space-world changes. Due to the lack of research, this research seeks to provide a thorough explanation of the overall situation of environmental change. For this reason, we applied some recognized and measurable methods to the collected atmospheric information to study the spatial variation of temperature and precipitation over the period 1901-2002. The method of approved environment change site is also used. Environmental changes focus on Rajasthan in this way to curb dynamic and acceptable risk mitigation systems. This section reviews the signs from the model study conducted on the variability of temperature and precipitation information over the last 40 years (1971-2010) at the above locations and the selection of remote locations. And is intended for consideration. Evaluations conducted by various experts are conducted in the near zone. Work relies on information obtained from the India Meteorological Department and the Irrigation Department. The default objective was to find a model/case of an array of temperatures and precipitation over the last 40 years. The rating was rehearsed for a large cluster in June and January, reflecting the summer and winter peaks in the rating area. The chart was created by recording state information from 12 scheduled meteorological stations at headquarters in each region and using a variety of data to capture a general model. The temperature design is investigated based on its three structures: annual average temperature, maximum average temperature, and minimum average temperature. On the other hand, the rainfall information collected by the Water Services Department reflects the normal year rainfall area and total area.

VARIABLE IMPACT ON DIFFERENT CROP

Very often, at the end of July or August, it rains enough to saturate the soil with profile (125 mm). These conditions have been shown to exist within 31 of 75 years. In addition, the beginning of the moon is generally delayed in this area (once every four years). Therefore, it is necessary to identify crops and varieties that can produce profitable crops under these conditions.

CROP PRODUCTIVITY

Many studies on the impact of environmental change on agriculture have reached a similar conclusion that environmental change reduces yields in the tropical region. Subsequent short-term environmental changes will almost certainly result in high-altitude returns throughout the longer evolving season, as indicated by the IPCC, but even limited warming is generally low-altitude. Decreases in the return of the geological disparities caused by environmental changes in the production of food are very important for the global placement of food. Food availability is a problem in tropical, given that about 800 million people are currently in poor health in the created world, even without the challenge to Earth. Crops are currently being developed in many parts of India and other states under the most extreme temperature obstacles, so the effects of unnatural weather changes will eliminate any movement. This is especially true in dry, unflooded areas where huge vulnerabilities exist. Similarly, moderate yields at 1 ° C for wheat and corn and 2 ° C for rice will significantly reduce yields in these areas. Most upbringing frameworks are considered regular, so they are prone to changes in the environment. Special climatic conditions are required in certain cases for optimal growth, harvesting, and further development of livestock.

Such an ideal period can change with temperature changes. The world's most vulnerable horticultural structures created are the dry, semi-dry, and dry sub-wet districts. The unpredictable and dry spell/flood continuous pattern of precipitation in these locations confuses crop yields, especially when yields occur in negligible terrain with little information. increase. Hot and colds cause physical damage to crops and destroy yields. Wounds are caused by high temperatures, leaf singing, and plant dryness in uncovered areas of the plant. Young saplings dry rapidly regularly as the soil temperature rises. In the lower range, elevated temperature accelerates the respiratory rate and unnecessarily contributes to the development of the next best.

IMPACT AND VULNERABILITY OF INDIAN AGRICULTURE TO CLIMATE CHANGE

So far, Indian agriculture seems to be a company that responds to expected environmental change ideas. Due to his discontinuity, he must create an instrument. 237 The work of the Indian Agricultural Research Council (ICAR) has just begun to determine the potential impacts of environmental changes on various crops, fisheries, and livestock. The following is a segment-specific survey.

GRAIN CROPS

The Asia-Pacific region will almost certainly face the most disastrous consequences for oat yields. Wheat, rice, and corn yields from the accident are estimated to be about half, 17% and 6%, apart from 205048. In any case, this yield disappointment will affect the food security of 4,444 people in South Asia, 1.6 billion people. The main reason for the decline in grain yields in much of South Asia is the expected rise in temperature from 0.5 ° C to 1.2 ° C. Wheat India is considered to be the second-largest wheat-producing country with a national wheat production of approximately 2708 kg/ha. Some of the major wheat-producing states are in northern India. Examples are Uttar Pradesh, Punjab, Hariyana, Uttarakhand, and Himachal Pradesh. Environmental changes can have a significant impact and increase wheat yields, as temperatures in Uttar Pradesh, Punjab, and Hariyana can rise by just 1 ° C. In February and March 2003, nighttime temperatures in Hariyana were 3 ° C higher than expected, reducing wheat production from 4106 kg/ha to 3937 kg/ha during this period. Assessing the impact of changes in properties on wheat production, the country's annual wheat yield can decrease by 6 million tonnes for every 1 ° C change in temperature. In any case, you can reduce the difficulty of your adaptation strategy, for example by changing the planting date or using different groups. It has been found that by changing certain agricultural practices, a 1 ° C rise can recover 44,433 million tonnes. The assessment also found that the impact of ecosystem migration on wheat production varies significantly from zone to zone.

MAIZE

Corn (Zea mays L.) is India's third most important grain, playing a huge role in food security, especially in mountainous and desert areas. Maize production in dry and semi-arid tropics is particularly sensitive to air conditions such as rainfall. Currently, precipitation is relatively exceptional during the southwestern and north-eastern storms, and in most cases, exceptional and lowest temperatures adversely affect corn harvests. In Tamil Nadu, assessments show that yields in 2020, 2050, and 2080 were reduced by 3.0%, 9.3%, and 18.3% from current yields. When it comes to creation, two huge advances are expected to occur. In any case, the yield of corn in the rainy season should decrease with increasing temperatures. In any case, this could be offset to some extent by the increase in precipitation. In addition, given rising temperatures, maize yields may decline throughout the winter in the central Indo-Ganges Plain and the southern part of the Plateau.

VEGETABLES AND LEGUMES

Most crops are exposed to harsh natural conditions, with irregular high temperatures and weight of soil moisture, which can reduce yields overall. Anyway, one query also shows that high CO2 concentrations can offset the adverse effects of high temperatures, especially for green vegetables that benefit from high photosynthesis.

CHICKPEAS

In the short term, vegetables and vegetables, especially crops, such as chickpeas, are better absorbed by the ability to utilize the additional photo, so carbon sink restrictions are very noticeable. Chickpea, developed with an increased CO2 load (up to 550 ppm), showed better performance in contrast to the plants developed with the current 370 ppm CO2 concentration. In certain cultivars, there was more pronounced shoot elongation and leaf elongation under increased CO2 and 18% increased seed mass. Anyway, elevated temperatures are most likely to reduce the beneficial effect of CO2 prolongation.

FOOD SECURITY STATUS

Food security needs to address both the physical and financial access to food. Livelihood security is not just an issue of livelihood, in any case, a progressively specific proportion of the section on good nutritional nutrition of families and people. Appropriately, sustenance security is researched nearby the tomahawks of straightforwardness, get to, and ingestion. The noteworthiness of favorable circumstances in sustenance security is other than underlined by Supreme Court's choices grasping food rights.

CONCLUSION

In India, there are as of now gauges of progress in the defencelessness profile because of environmental change. Drought is the most critical consequence of an environmental change in poor India. Bone-dry Rajasthan is described by unpredictable precipitation with conflicting dissemination, high diurnal, and occasional temperature, high daylight, and serious breeze conditions. Planting precipitation has additionally happened in some cases in the eastern piece of 1 June and in the west piece of 2 June.

This early or poor start of planting precipitation adds to the noteworthy variety in crop efficiency a seemingly endless amount of time after year. The physical effect of environmental change is normal in India, all in all, be considered a 2-4 degree C ascend in normal surface temperature and Increases in both storm and non-rainstorm precipitation months. 15 blustery days, an expansion in precipitation power of 1-4 mm for each day, and an ascent in precipitation force by a reduction of more than 15 days. in India.

Indian farming is twice helpless because of environmental change. It is from the start the most defenseless against the impacts of environmental change on lunars because about 60 % of the all-out horticultural region in India is taken care of with downpour. Second, more than 80 % of ranchers in India are minimal hence less fit for taking care of the effects of environmental change on farming. India's 200 reverse areas are recognized for the huge rainfed horticulture action by the Planning Commission.

In many pieces of the nation, ranchers favor wheat or rice with changing nourishment propensities and economic situations. The environmental change is anticipated to have genuine results, especially for wheat, for these significant harvests. The elevated level of subsistence cultivating with little land possessions, most of the agrarian land, which could impact/increment his powerlessness to environmental change, is downpour taken care of Fair climate marvels, for example, dry season and twisters, are normal, and enormous varieties in ranch efficiency the nation over. Nourishment insurance is the capacity of nourishment handling forms over the whole evolved way of life.

Nourishment security will affect sanitation on all components of the global, national, and neighborhood frameworks of nourishment creation that are relied upon to affect every one of the four elements of nourishment security. Late gauges for nourishment creation show a multiplying between the normal populace and monetary development, extending from 2 billion to 4 billion tons of grain each year. India would be genuinely influenced by environmental change and atmosphere fluctuation. Expanded recurrence and examples in extraordinary conditions may influence dependability and access to nourishment supply.

The expansion in crop misfortunes because of these outrageous occasions overpowers the beneficial outcomes of moderate temperature rises. Woodland-based natural surroundings may potentially affect the nourishment creation of biological system benefits by expanding the danger of backwood fires, bug spray flare-ups, and other timberland aggravations. As of not long ago, most environmental change sway evaluations on the nourishment and yield area have focused on the impact on crop creation, considering different components of the nourishment framework and the interconnections between the distinctive nourishment creation frameworks. The yield design is regularly connected to the accessibility of creature grub, a significant nourishment security segment in India and other South Asian nations.

The environmental change would require more noteworthy protection from and adjustment to the effect of environmental change among ranchers and country networks. The key to actualizing reasonable and centered procedures is understanding the effect of environmental change in different agro-climatic areas. The effect of environmental change on conduct, while its financial outcomes are progressively being felt, remains to a great extent muddled. A meta-investigation of harvest execution tests shows that a potential increment in the photosynthetic action of plants with expanded CO_2 fixation is achievable just when associated with the ideal temperature but in future climatic conditions the temperature and precipitation levels are relied upon to change erratically.

The assessment of these various parameters offers a superior comprehension of the worldwide dangers to agribusiness in the environmental change time.

<u>REFERENCES</u>

1. Abeysingha NS, Singh M and Islam A 2016. Climate change impacts on irrigated rice and wheat production in GomtiRiver basin of India: a case study. SpringerPlus 5, 1250

2. CAPRI 2019.Vision 2030, Central Arid Zone ResearchInstitute, Jodhpur, India.

3. Cline W. 2007. Global Warming and Agriculture: Impact Estimates by Country. Peterson Institute of InternationalStudies Washington D.C, U.S.A.

4. Census 2011.Census 2011. Government of India, New Delhi.

5. CGWB 2017. Dynamic groundwater resources of India 2017.Department of water resources, Rd & GR, Central Groundwater Board, Government of India.

6. Darwin R F, Lewandroski J, McDonald B and Tsigas M. 1994.Global Climate Change: Analyzing Environmental Issues and agricultural Trade within a Global Context. In: Sullivan J. (ed)Environmental Policies: Implications for Agricultural Trade, Foreign Agricultural Economic Report No. 252, United StatesDepartment of Agriculture, Economic Research Service, Washington, D.C., U.S.A.

7. Deschenes Oand Greenstone M. 2007. The Economic Impact of Climate Change: Evidence from Agricultural Output andRandom Fluctuation in Weather. American Economic Review97(1): 21-53.

8. GoI 2013a. Twelfth Five Year Plan (2012-2017). the planning commission, Government of India, New Delhi.
(3) (PDF) Assessment of Climate change impact on wheat yield in Western Dry Region: A District level analysis. Available from:
https://www.researchgate.net/publication/348237129_Assessment_of_Climate_change_impact_on_wheat_yield_in_Western_Dry_Region_A_District_level_analysis [accessed Jul 20 2022].

YOUR KNOWLEDGE HAS VALUE

- We will publish your bachelor's and
  master's thesis, essays and papers

- Your own eBook and book -
  sold worldwide in all relevant shops

- Earn money with each sale

Upload your text at www.GRIN.com
and publish for free